COLLINS

do it

SERIES

WIRE IT

Robert Henley

Series Consultant Editor: Bob Tattersall

CONTENTS

HarperCollins*Publishers*

Introduction

Anything to do with electricity is thought to be difficult and dangerous. The usual advice is 'leave it to a qualified electrician'—and this attitude is encouraged by all competent authorities. The truth is, however, that many electrical jobs are well within the competence of the amateur who takes the trouble to understand the subject and who takes obvious precautions, such as first switching off at the mains before starting any work.

Read this book carefully and try to understand the basics of home wiring before planning any changes. If you find yourself still baffled, leave the job to a competent electrician. Do not try to tackle anything like a complete home wiring until you have first successfully completed some of the simpler jobs.

Care and accuracy are most important. 'It looks right' or 'that will do for now' might be acceptable for some jobs around the home, but this certainly does not apply to electricity. Everything must be exactly right. You must check everything carefully before completion and switching on.

There are regulations about the supply of electricity to the home. The electricity board can refuse to connect your home to their supply if they consider the wiring to be unsafe but, in practice, their interest is limited to making sure you have an effective earth system so that, if there should be a fault, your wiring is not likely to jeopardise their equipment or the supply of electricity to other homes.

There are also rules for electric wiring safety in the home. These are contained in the *Regulations for the Electrical Equipment of Buildings* (fifteenth edition), published by the Institution of Electrical Engineers, and better known to all professional electricians who follow them as 'the IEE Wiring Regs'. Following 'the Regs', a copy of which can be found in any reference library, is not only advisable in the interest of your own safety but it also means the work will satisfy the electricity board.

The Regulations are not too easy to understand—even for the professional—but we have tried here to simplify them, as they apply to home wiring.

Left *A peaceful monochromatic colour scheme has been brought alive by touches of yellow and good lighting. If you don't want to embark upon the individual wiring of ceiling lights, you could use a track system connected to the* ceiling rose. *You can then have as many fittings as you want.*

Above *In planning this galley-type kitchen, it was essential to have sufficient outlets for all-electric appliances.*

ELECTRICITY IN THE HOME

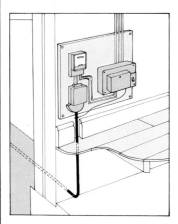

The Meter

Electricity comes into the home through a thick, armoured, two-core service cable made up of the live and neutral circuits. The cable goes to the meter through a box containing a fuse (service fuse). This is sealed by the electricity board and must never be touched. If this fuse fails call the emergency service of the electricity board (it is in the telephone directory but it's a good idea to write the number on the wall, close to the meter). Check all the other fuses in the house first, as it is very rare for this fuse to 'blow'. Check with neighbours that it isn't a power cut.

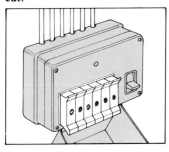

The Mains Switch

From the electricity meter the electricity supply goes to the mains switch which controls the electricity supply to the house. In houses 30 or more years old it is still common to find the mains switch in a separate box. In modern homes it is part of the fuse box, where the main circuit is split into a number of separate circuits — power, lighting, cooking and so on — each with its own fuse.

The electrical equipment and wiring from the meter onwards is the responsibility of the householder. The householder also has to make sure there is an effective earth wiring point to which all the other wires in the home are connected.

Modern wiring practice for effective earthing is to fit an electro-mechanical earth—a *residual current circuit breaker (RCCB)*. These are also known as *earth leakage circuit breakers (ELCB)*.

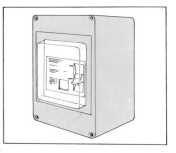

An **RCCB** is designed to keep a constant check on the live and neutral circuits, which normally operate in equal balance. If the

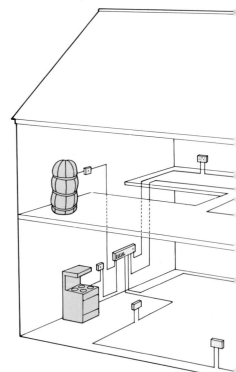

RCCB notices a sudden change in the current flow, it instantly and automatically switches off the current. RCCBs are sometimes incorporated with the mains switch in the consumer unit. More often they are used as separate units to protect one circuit—especially the one for outdoor sockets or switches where the risk of electric shock is always greater.

From the *fuse box* (sometimes called a *consumer unit*), cables form separate circuits, each containing a *live, neutral,* and *earth* wire. In these cables the *live* is coloured *red* and the *neutral, black* (flexes have different colours — see p. 8). The earth wire is usually bare.

All three wires are encased in a grey or white insulating material. In most homes the cables are imbedded in the plaster on the walls or run in steel, aluminium or plastic tubes. Older homes may have cloth or rubber covered cables, black or red.

The Ring Circuit

The circuit supplying the socket outlets and plugs is the ring circuit. The ring circuit has the 13 amp socket (with square holes) to fit a square, three-pin plug that can supply an appliance up to 3000 watts. This type of socket has safety shutters that close automatically when the plug is withdrawn so children can't poke anything in to touch the live terminals.

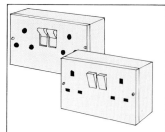

Modern homes use the ring circuit system of wiring for sockets that have shuttered square entrance holes for the matching square pin plugs. Earlier systems use round pin plugs and sockets. These systems should be replaced.

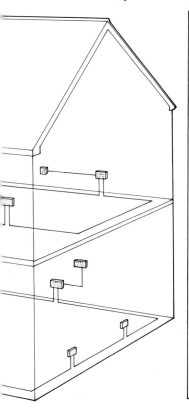

Tools

There are only a few extra tools, in addition to normal DIY items, that are needed for home wiring jobs.

Wire cutters (or diagonal cutters) for cutting cable and flex to length.

Wire strippers are used to remove the insulation from individual wires, not the outer sheath of cable. There are several types available and they adjust to accommodate different thicknesses of wire.

A *sharp knife* is necessary to score the outer cable sheath. Be careful not to cut through the insulation of the wires inside.

A *thin-bladed screwdriver* is best for the brass terminal screws in many electrical fittings.

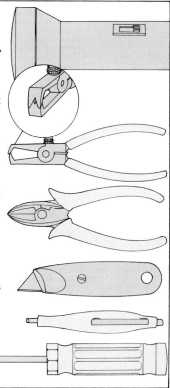

Fuses

The cartridge fuse in the 13 amp plug is designed to protect the flex by being a deliberate weak spot in the electrical circuit. If more current flows than the flex is designed to carry, it could get dangerously hot. To prevent this, all fuses are made of weaker material than the rest of the circuit, so that it melts (or 'blows') when overloading occurs. This stops the flow of current and prevents further trouble.

Main Fuses

The fuses in the main fuse box (consumer unit) operate on the same principle. There is one for each circuit, eg lighting, power sockets, cooker etc.

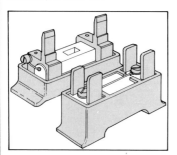

The most common form of fuse for the consumer unit has a fuse wire held in position by brass screws.

A modern alternative to rewirable fuses are *Miniature Circuit Breakers (MCBs)*. These look like ordinary switches or push buttons and they automatically flick themselves off if the circuit they're protecting is overloaded. The

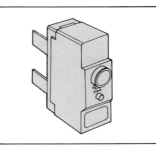

circuit can be brought back into use simply by pressing the switch down (or pushing the button in) to reset the circuit breaker.

Do not get confused between an MCB and an RCCB. The MCB operates only in the case of overload (too much current), while the RCCB operates in the event of current flowing to earth (out of the circuit—perhaps through a faulty appliance).

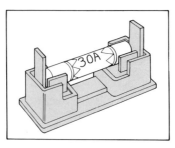

Another type of fuse for consumer units is a *cartridge fuse*. This is similar to the cartridge fuse in the 13 amp plug and may be replaced as easily.

Replacing a Rewirable Fuse

Get a torch (keep one near the fuse box). Switch off the main switch. If possible, find out the cause of the fuse blowing. Switch off the faulty appliance or lamp and unplug it, otherwise the new fuse may blow.

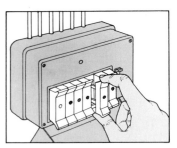

Find the blown fuse by taking out and inspecting each one in turn. It is usually obvious which one has blown — not only will the fuse wire be broken, but often there will be scorch marks around the fuse carrier.

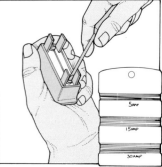

Replace the old wire with a fresh piece of the correct amp rating. Don't stretch or strain it when tightening the screws. Replace the fuse carrier, close the box, and then turn on the switch. If the fuse blows again, or if there is any doubt, send for an electrician.

Tips

You can make the job of finding blown fuses easier by labelling each fuse in the consumer unit. That way you can see quickly which fuse belongs to which circuit, eg upstairs lighting, ground floor power, etc.

Reading the meter

How Electricity is Measured

The 'flow' of electricity is measured in amperes (amps or A); the 'pressure' is measured in volts (V), and its power (the work it can do) in watts (W). The standard voltage in Britain is 240 V (Northern Ireland 230/240V). All fuses and fuse wires are marked in amps.

1000 watts used continuously for one hour is equivalent to one kilowatt hour (*kWh*). This is the *unit* by which electricity is measured and is stated on the bill.

A metal rectangle, called a rating plate, is usually found at the rear of an appliance. It shows the maker's name, the model number (which is quoted for servicing) and the amperage, the voltage and, usually, the wattage. Multiplying the amps by the volts will give the watts $(A \times V = W)$ — the amount of power the appliance uses, and so a guide to the running cost.

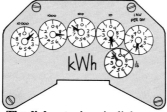

The dial meter has six dials. Each pointer goes round in the opposite direction to its neighbour. The dials record units: tens of thousands, thousands, hundreds, tens, and singles. Ignore any dials that register tenths or hundredths of a unit. These are usually coloured red.

Start by reading the dial showing single units, and write down the figure. Then read the dial showing tens of units, then the one showing hundreds, then thousands and then tens of thousands, working from right to left and writing them down in that order. Always write down the number the pointer has passed (this isn't necessarily the nearest number to the pointer). So, if the pointer is anywhere between 3 and 4, write down 3. If the pointer appears to be directly over a figure, say 7, look at the pointer on the dial immediately to the right. If this pointer is between 9 and 0, write down 6. If it is between 0 and 1, however, write down 7.

This is the *digital meter* that shows units of electricity by a simple row of figures. Subtract the previous reading from the new reading to see the number of units (kWh) used.

Running Costs

The Major Popular Appliances

Cooker Uses about 4 units a day cooking for a family of four.
Dishwasher Washes a family's dinner dishes for about 3 units.
Freezer Uses about 1.5–2 units per cu ft per week.
Electric fires and fan heaters With a loading of 2 kw, 2 units per hour.
Refrigerator The table-top height size uses about 1 unit per day; the larger sizes 1.5 units per day.
Shower 1 unit for 2 showers.
Iron Irons for over two hours for 1 unit.
Spin dryer Spins about five weeks' laundry for 1 unit.
Tumble dryer Uses about 2 units for one hour.
Automatic washing machine About 2 units of electricity for one wash load on prewash and hot wash.

Smaller Appliances	For 1 unit of electricity you can:
Air conditioner	run for 1 hour in summer
Blanket (over)	all night for 2 or 3 nights
Blanket (under)	use every evening for a week
Blender	make 500 pints of soup
Coffee percolator	make about 75 cups of coffee
Floor polisher	polish for 2½ hours
Food mixer (stand model)	mix 67 cakes
Hair dryer	use it for 3 hours
Health lamp (infra red and uv)	use it for about 4 hours
Kettle	boil about 12 pints of water
Lawn mower	do 4 hours' grass cutting
Power drill	have about 4 hours' drilling
Sewing machine	sew 11 childrens dresses
Shaver	have over 1,800 shaves
Television (colour)	have about 5 hours' viewing
Toaster	make 70 slices of toast
Vacuum cleaner	clean for about 2 hours
Video recorder	have about 10 hours' use
Waste disposal unit	grind about 1 cwt of rubbish

CABLES & FLEXES

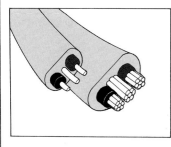

Cables

The supply of electricity depends on cables and flexes. *Cables* are oval and are meant to be permanently fixed, held flat by special clips. Cables are made of three conductors (wires). One covered with *red* insulation (the *live,* one covered with *black (neutral)* and a bare wire called the *earth continuity conductor.*

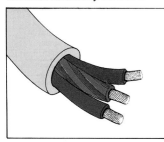

Flexes

Cables have relatively few, but thick, strands of wire while *flexes* are made of many finer strands to give them the flexibility needed to run between plug and appliance. The insulation on the three conductors is coloured differently from cables. The *live* is *brown;* the *neutral* is *blue;* and the *earth* is *green yellow.* The outside insulation of most three-core flex is plastic but it may be rubber with a braided linen cover for appliances such as irons. Also, some heaters and immersion heaters require a heat-resistant insulation.

Double-insulated appliances only need a *two-core flex.* This is because they are made in such a way that no current could leak to the outside of them and cause electric shock to the user. Double insulated appliances all bear a symbol of one square inside another.

Two-core flex is also used for standard and table lamps unless they are made of metal.

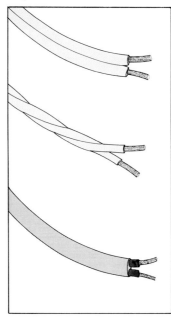

There are three types of *two-core flex. Parallel twin flex* has two wires, each in its own insulation cover, joined together and not surrounded by an overall covering. Although the insulation on both wires is the same colour, there is usually a small rib along the side of one wire. If possible, you should use this wire as the live conductor.

Twisted flex has a braided fabric cover which makes it better for use with table lamps.

Finally, *two-core sheathed flex* with colour-coded insulation is made in different forms and amperages.

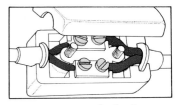

Never try to extend a flex by joining on an extra length using insulating tape. Purpose-made flex connectors in tough plastic are sold, but it is usually cheaper to buy a complete length of flex and connect it to the appliance.

Avoid long flexes as it is all too easy to trip over them. It is even worse if they have to be run under a carpet where they can be damaged and may even cause a fire. The best solution is to instal extra sockets (see p10) instead.

When fitting a flex to a plug or appliance, only remove enough insulation to enable the wires to be connected to the terminals. There should never be any bare wire exposed. Also, never stretch the insulation, and be sure to use any flex grip in the plug or appliance so that an accidental tug on the flex will not pull out the conductors.

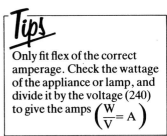

Tips

Only fit flex of the correct amperage. Check the wattage of the appliance or lamp, and divide it by the voltage (240) to give the amps $\left(\dfrac{W}{V} = A\right)$

Wiring a 13 amp plug

Ring Circuit

All modern homes have their socket outlets wired on ring circuits or loops of 2.5 sq mm twin-core and earth PVC sheathed cable that starts at the consumer unit, 'visits' each socket in turn and returns to the consumer unit. Each ring is protected by a 30 amp fuse. Each ring circuit is limited to supplying socket outlets and fixed appliances over an area of 100 square metres, but the number of sockets on a ring is unlimited. Because the maximum load that can be taken at one time is about 7200 watts, it is usual to have two ring circuits, one concentrated near the kitchen area.

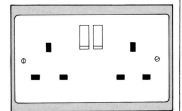

It is best to install double sockets rather than single, and they should have switches.

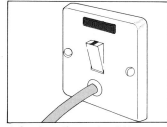

A fixed appliance should be supplied from a *fused connection unit (FCU)* rather than a socket. Appliances such as freezers should be connected by a plug and socket (or FCU) fitted with a red neon indicator to show that the electricity supply is connected and working.

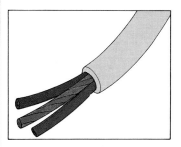

Cut away about 50 mm of the sheath on the flex, without cutting the insulation on the individual wires.

Open the plug by undoing the screw, and remove the cartridge fuse. Cut the wires off so they are just the right length to go through the fixing holes or around the clamping screws. Strip off just enough insulation and no more.

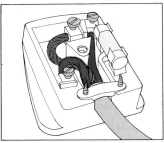

The *green or green yellow* is the *earth wire,* and goes to the larger of the three pins—marked 'E' or 1. The *brown lead* goes to the pin marked 'L' (for *live*). The *light*

blue lead goes to the pin marked 'N' (for *neutral*). A two-core flex can be connected to a three-pin plug, but leave the earth pin unconnected.

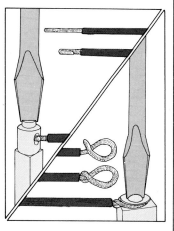

Clamp the wire ends firmly in the terminal connections, as shown. The outer sheath of the flex must be gripped firmly under the clamp where it enters the plug. This prevents the wires pulling out of the terminals.

Replace the cartridge fuse. Most plugs are supplied with a 13 amp (brown) fuse, but a 3 amp (red) fuse should be used for lamps and appliances rated up to 720 watts. See the appliance rating plate or the maker's instructions, if you're in doubt about which fuse to use.

Before screwing the cover on to the plug, make sure there are no stray strands of bare wire and that the terminals are tightened hard down—not just finger tight.

Replace the plug cover and screw it into place. There should be no looseness or rattles of any kind when you've finished, and all screws or nuts holding the plug together should be firmly tightened.

Adding extra sockets

Extra sockets can be added to the ring circuit. These are added on a *branch line* from the ring called a *spur*. The spur is wired from the back of an existing socket and can supply two singles or one double socket. Sockets should be placed not less than 150 mm from the floor.

To connect a spur to the ring, turn off the power at the mains switch and ease an existing socket from the wall by undoing the screws in the faceplate.

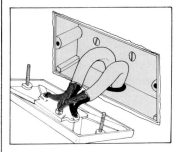

Behind the faceplate is the *mounting box* containing up to three cables. Check whether this socket is on a ring or on a spur. (Leave spur sockets alone.) *One-cable* sockets are always spurs; *three-cable* sockets show that a spur has already been added to the ring and these also should be avoided. *Two-cable* sockets are probably on a ring, but they just might already be part of a spur, so it is a good idea to check with a **circuit tester** (see page 11).

Next, decide where you will run the cable. In many homes you can run the cable under the floor but, where there is a solid floor, the cable will have to run in the plasterwork of the wall. Alternatively, you can use hollow metal or plastic skirting to conceal the cable. The cable could also be run along the surface of the skirting, but it

should always be covered in plastic conduit to protect the cable and also to disguise it.

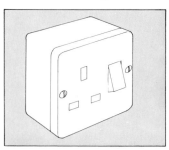

The new socket can be mounted on a plastic surface box, but flush-fitted metal (or plastic) boxes are the neatest.

Cutting the plaster to hide the cable is not as difficult as it might seem, especially if redecoration of the wall is planned.

Using a sharp knife, score two lines 25 mm apart along the proposed route. Use a club hammer and a brick bolster to chip out enough plaster to allow the cable to fit comfortably.

Another method is to use a plaster router bit fitted to an electric drill or router, after first drilling a series of holes along the proposed route. Check carefully that you are not going to interfere with or damage any cables already in the wall.

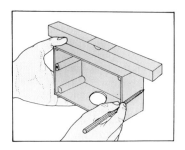

The hole to take the metal box usually needs to be cut deeper than the depth of the plaster. First place the box against the wall and trace the outline with a pencil.

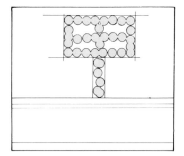

Use a masonry bit (marked with tape to a level a little deeper than the box) to cut a series of holes along the lines, and then drill holes over the whole area.

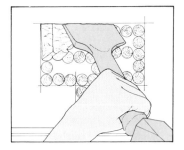

Now the bolster will remove the plaster and brick easily. Check that the mounting box fits and then knock out the required holes in the box for the cable.

Always fit a *grommet* (a rubber ring) into each hole to prevent the cable from rubbing against the metal. Place the box in the hole and mark the wall for a screw, take out the box, drill a hole and insert a wallplug. Finally, fix the box, making sure it sits level—a crooked socket always looks untidy.

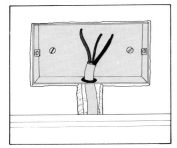

Take the end of the length of cable (2.5 sq mm PVC sheathed twin-core and earth) through the new box's knockout hole and leave about 75 mm protruding from the box.

Now run the cable to the existing socket, either in a channel in the plaster or under the floor. On a ground floor the cable can normally be placed in the void beneath and clipped to the joists.

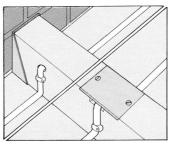

On an upper storey, where the cable runs at right angles to the joists, pass it through holes drilled at least 50 mm below the joist's top edges. Otherwise, you could cut notches in the top edges of the joists to take the cable, but stiff metal plates must cover the cable and be fixed to the joist.

Take the cable to the socket on the ring where it is to be connected. Remove this socket from the cable and unscrew the box. Now complete your channel to give access for the new cable. This will probably mean knocking out another

circular metal blank in the box and fitting another grommet. Pass the cable through the grommet and screw back the box. Cut the new cable to leave about 75 mm protruding.

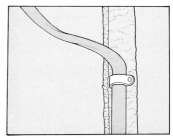

Clip the cable along its length where necessary, to hold it in place. There are clips made for the purpose—be sure to use clips of the correct size and shape for your cable.

Now mix your plaster, fill in the channel and around the new box, and leave to dry.

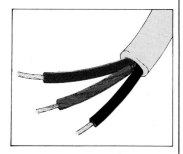

Next prepare your cable ends, stripping away the outer insulation with a sharp knife. Take the red and black covered wires and remove 12 mm of insulation from each with the wire stripper. Cover the bare copper earth wire, where exposed, with green or green and yellow PVC sleeving.

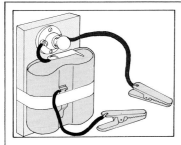

A circuit tester can be made from a 9 volt battery, some bell wire, a pair of crocodile clips and a miniature bulb and bulb holder. Use the tester to link the cables' red wires. If

the bulb lights, the socket is on a ring circuit.

You can also use the circuit tester to do an earth test. Place one clip on the metal part of the appliance, and the other on the earth pin of the plug on the appliance (this won't work on double insulated appliances). If the bulb is bright, the earth is safe; if it's dim, the earth is weak and should be examined. If the bulb lights when the clip is placed on either of the other two pins, there is a serious fault.

Cookers

In the existing box, the three red, three black, and three copper (earth) wires are matched and each group of three connected to the appropriate terminal on the socket (see illustration).

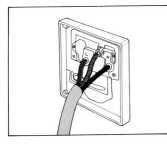

In the new box, one wire is connected to each terminal (see illustration). Screw each terminal tight, making sure all the wires are fastened under the screws.

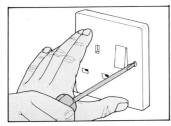

Replace the sockets by holding the faceplate against the box and pressing the cables in carefully. Fix the sockets to the boxes with the bolts provided, switch on, and test.

Double Socket for a Single

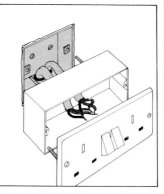

It is sometimes possible to fit a new double socket on to an existing flush box using a slim surface mounting *pattress*. This is about 19 mm deep and fits over the existing box. There is a mounting hole at each side so that it can be screwed to the wall. The double socket is then fixed on to the pattress in the same way as onto a surface mounted box.

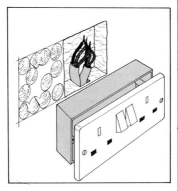

The more usual method is to remove the old mounting box and cut out a hole to take the new mounting box. The method of cutting the hole, mounting the new box and connecting the socket is the same as described in *Adding Extra Sockets* (p10).

Fitting an Electric Cooker

The cooker must have its own circuit—30 amps for cookers up to 12 kw and 45 amps for those above 12 kw, such as the large split-level ovens and hobs operated on the same circuit.

The cable is run from the consumer unit to the cooker control unit mounted on the kitchen wall.

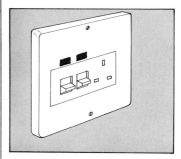

The cooker control unit has a *double pole switch*—one that controls both the live and neutral conductors.

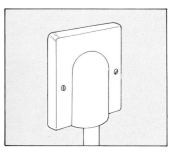

With a free-standing cooker, it is usual to fit a *terminal outlet box* about 600 mm from the floor. This is joined to the control unit with a second length of cable. Finally, a length of flex joins the cooker to the terminal outlet box, allowing the cooker to be drawn away from the wall for cleaning.

A split-level cooker can be controlled from one unit, but neither the hob nor the oven must be more than 2m from the switch. The cable supplying both is divided in the cooker terminal, or else taken first to one unit and then on to the other.

Plan the route for the cable either under the floor or through the kitchen ceiling to above the consumer unit. Channel out the walls to take the cable. The 6 sq mm or 10 sq mm cable is heavier and thicker than the 2.5 sq mm used for the ring circuit, so a deep channel will be needed. If you choose to run the cable on the surface, use cable clips every 200 mm or else conceal it in plastic channelling or conduit.

Open up the floor or ceiling space and cut holes in the joists to take the cable if it has to pass across them.

Mark the position of the metal *mounting box*. If it is to be recessed, remove a portion of the wall to take the box. Knock out the blanks for the cable entry holes and fit grommets into the holes. Screw the box to the wall.

Insert the cable into the box and strip off about 200 mm of outer cable sheath. Fit a length of green yellow sleeve on the earth wire. Remove enough of the insulation on the live and neutral wires so that they can be connected to the terminals on the unit. Repeat this operation with the cable to the cooker. With a split-level cooker there may be two cables to connect to the outlet terminals of the unit. Make sure the red wires go to the live and the black to neutral. The terminal screws must be tightened to hold the cable ends effectively. There should be no

loose strands of wire. Check that the earth connections are tight.

Fit the unit into its box and fix the retaining screws.

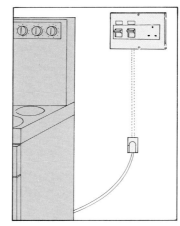

Free-standing cookers
The terminal outlet unit for the free-standing cooker has a metal frame with a terminal block that fits into a mounting box. This box must be sunk into the wall. Channel the cable into the wall up to the cooker control unit. Complete any replastering.

Prepare the ends of the cable in the same way as on the control unit (see p12) and connect to the terminal block. Fit the terminal block on the mounting box followed by the front plate cover.

Now the other end of the cable has to be connected to the consumer unit and this is a job best left to a professional electrician or the electricity board. You will have to arrange a temporary disconnection. If there is no spare way to the consumer unit, a separate double pole switch unit with a 30 or 45 amp fuse will have to be installed. The unit is mounted next to the consumer unit and needs two 3 m lengths of 16 sq

mm, one red, one black and a 6 sq mm earth cable of the same length for connection to the terminal block by the electricity board.

To install a free-standing cooker, first find the *cable connection point* at the rear of the cooker. This is usually located behind a metal plate with a cable clamp. The cable connecting the cooker to the terminal outlet box should be of the same size and capacity as the circuit cable. About 2 m should be sufficient.

Prepare both cable ends for connection, leaving more bare wire at the cooker as it usually has to be wound around a screw and clamped down with brass washers and screws.

Switch off at the mains and at the cooker control box. Open the terminal and make the connections. Replace the terminal cover, switch on and test.

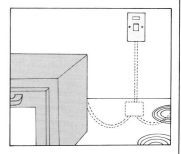

To install a split-level cooker, the cable connection is taken from the cooker control unit through a flush mounting placed behind the hob and oven. The cable is clamped inside the mounting box and a terminal outlet front plate fitted. Connection is then made to the oven and hob at the terminals provided.

IMPROVING YOUR LIGHTING

There are plenty of good ideas that will help you to improve your lighting, and many need little or no structural work or rewiring. Changing a light switch to a *dimmer*, for instance, allows you to create mood lighting and is a job that takes only a few a minutes. For any of these jobs, though, switch off at the mains first — not just the switch at the door — or else remove the lighting circuit fuses from the consumer unit.

Light Fittings

Rise and fall ceiling roses allow the height of the shade to be adjusted simply by finger touch. A circular box with a spring balance replaces the ceiling rose. The fitting should suit the weight of the shade.

Lighting track is one of the best ways of getting more light without rewiring or tearing into walls and ceilings. The track can be mounted on to a wall or ceiling and a variety of small, neat light fittings can be clipped into the track anywhere along its length. It is particularly useful for mounting small spotlights that concentrate their light on to nearby walls, furniture or pictures.

Spotlights produce a beam of light of different sizes and intensity. One of the best methods of using them is to conceal them in the ceiling so that only the bulb can be seen.

The ceiling space in most homes is sufficient to take this kind of fitting. The average recessed depth is usually less than 125 mm. A circular hole must be cut in the ceiling and the cable connection can be taken through the space between the joists to the nearest lighting point. The hole should be cut carefully, although any rough edge will be hidden by the trim on the fitting.

As many as four of these fittings can be run from an existing centre ceiling point, the four cables being connected together within the ceiling rose or a new junction box. It is a good idea to control all these with a dimmer.

The *'eyeball'* is the most versatile of the recessed fittings. The housing that holds the

A row of recessed spots give directional lighting, making this corner ideal for sewing or reading, whilst not over-illuminating the whole room. They are easily fitted, provided you have sufficient height in the ceiling space; check this before buying your fittings.

lamp swivels to direct light at any angle. The smallest eyeball (for mains voltage) needs a hole of only 100 mm diameter and a depth of 115 mm. Eyeballs that take a larger lamp need holes of 200 mm and the recessed depth is about the same.

Cones, tubes and squares create pools of light of different sizes, depending on design. Some can be adjusted or only partly recessed in the ceiling, making the fitting suitable for shallow ceilings. One version has the lamp set at an angle to throw light sideways on to the wall, like the eyeball.

Circular glass fittings have a lens that hides the lamp and spreads the light. Some fit flush

This large basement kitchen/dining room relies on a well thought out plan that provides light only where it's wanted, by means of strategically placed spots. The fitting over the dining table is separately switched.

with the ceiling, some are slightly recessed (which tends to narrow the spread of light) and others have the glass and trim surface mounted. Clear, milk and sculptured glass covers are available and most are designed for the tungsten filament (GLS) lamp. These light fittings are suitable for bathrooms and kitchens.

Lamps are the first essential of any lighting. Popularly called 'bulbs', there is a wide choice.

Pear-shaped bulbs are known in the trade as GLS (general lighting service). Pearl coated and white (argenta) give a diffused light, while clear are for clear and tinted fittings where the undiffused light adds a sparkle. 60 and 100 watts are the most useful sizes. Long-life bulbs give a little less light and cost a few pence more, but they last for 2000 hours instead of the usual 1000.

Most bulbs have bayonet cap fittings (BC), but many continental fittings use screw cap (ES) bulbs.

Decorative bulbs are variations on the pear shape, usually for situations where the bulb is visible. Candle flame shapes are for wall lamps. Coloured bulbs are for parties and at Christmas, and those with pink pearl coating give soft background lighting or create a welcoming glow in the hall or on the patio. There are also specially shaped bulbs with the white (argenta) coating. Some of the decorative bulbs have cap fittings that are smaller than normal — called SBC or SES.

Crown silvered bulbs have the front silvered so the light is then thrown backwards against a reflector, to give a glare-free spotlight beam. These are used in special fittings, some of which need special small bulbs — 40, 60 or 100 watts.

Internally silvered bulbs (sometimes called reflector lamps) are silvered at the back so all the light is reflected forward (like a spotlight). There is a choice of narrow or wide beam and different sizes to suit special fittings — 40, 60 or 100 watts.

Pressed glass (PAR38) are strong spotlights made of toughened glass and are capable of being used outdoors as well as indoors. They are usually ES (screw cap) and need a special fitting. These bulbs have a long life — 5000 hours — and are made in 100 and 150 watt sizes.

Linear filament bulbs are tube-shaped ordinary bulbs (not to be confused with fluorescent tubes). They are usually used in wardrobes and under worktops and may have double-end fittings or single, to match the fitting — 35, 40 or 75 watts.

Fluorescent lamps give five times as much light for the same number of watts as filament lamps. They cost more, but they last much longer — about 7500 hours. Their sizes range from miniature tubes, 300 mm long (only 8 watts) to the most popular home sizes — 1200 mm (40 watts) and 1500 mm (65 watts). The colour of the light they produce is called 'warm white' but you can ask for de luxe warm white (sometimes called by a brand name such as Homelite). The colour of the light from these is more like the light from a filament bulb and not the rather white, cold tone of most fluorescents.

Almost all tubes can be disguised with a *baffle*. There are various translucent plastic covers for ceiling use in the kitchen. There

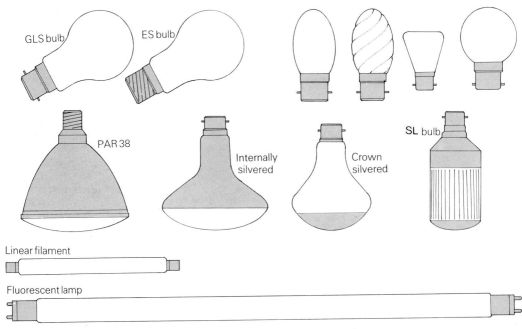

GLS bulb

ES bulb

PAR 38

Internally silvered

Crown silvered

SL bulb

Linear filament

Fluorescent lamp

are also circular tubes with covers for use in the centre of the ceiling in a living room or bedroom.

New thin fluorescent tubes that are twisted into a compact shape and fitted into a glass jar are threatening to replace conventional lamps. The smallest of these uses only 9 watts of power, but gives the light of a 40 watt GLS bulb, and the 18 watt version gives the same light as a 75 watt bulb. These new lamps fit into conventional bulb holders and, although they are far more expensive to buy, they last for 5000 hours and offer very reduced running costs. There is also a type — the 2D — that has a futuristic shape.

Dimmers vary the amount of light given out by a bulb. Most dimmers can be fitted in place of an existing switch, or there are free-standing models for table lamps. Several dimmers can be mounted together on a twin-switched plate. Always suit the loading of the dimmer to the lighting load (see maker's instructions). Fluorescent lamps need special dimmers. Connection of a dimmer switch is the same as for a rocker switch.

Other Devices

Pullcords are needed in a bathroom for safety; they are also useful in a bedroom where the pullcord can hang down close to the pillow, making it easier for the elderly and invalids.

Timers are usually plug-in and are useful for turning equipment (such as a table or standard lamp) on and off when you are away from home (an anti-burglar device).

A Besa box is a circular mounting box for some pendant wall and porch lights that have a circular base. They are recessed into the wall in the same way as square boxes.

Terminal blocks, made of white or black plastic, have pairs of screws for connecting cables inside mounting boxes. They are purchased in lengths and cut off in strips of two or three. They are often used to connect light fittings to the cable inside the Besa box.

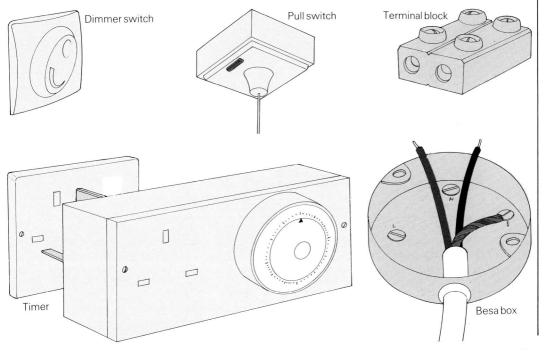

Dimmer switch

Pull switch

Terminal block

Timer

Besa box

Fitting ceiling roses & holders

Before beginning any work on ceiling roses or lamp holders, be sure to switch off the electricity at the mains switch—it is not enough to switch off just the light switch.

Lamp Holders

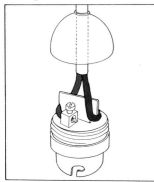

A lamp holder is needed for all lamps. To fit a new one, first unscrew the cover and slacken the flex from around clamping grooves on the centre pillar. Unscrew the terminals and remove the flex ends. If the flex is rubber insulated and cotton braided and has been in use for some time, it will almost certainly require replacing with twin-core, heat-resistant flex.

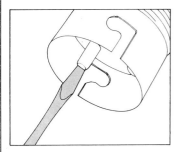

Push each terminal plunger up and down to see if the tension is still good. If not, or if the moulding is damaged, replace it with a new lamp holder. Thread a new cover on to the flex and remove the insulation from the ends of the wires, twisting each bare end neatly to keep the strands together.

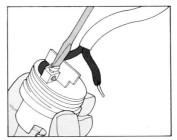

Place the flex ends into the terminals and tighten firmly. (If a flex with brown and blue conductors is being used it will not matter which terminals are chosen.) Place the flex in around the pillar in the grooves provided. Screw down the cover, keeping the flex slack.

Ceiling Roses

A ceiling rose connects pendant (hanging) flex to mains. Multi-outlet versions allow you to suspend up to five pendant fittings from one ceiling outlet. They are useful in a stair well or over a coffee table in a corner; or the flexes can be looped through hooks to spread light over the ceiling.

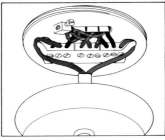

To replace a ceiling rose, switch off at the mains switch. Undo the cover of the rose. If it is an old porcelain pattern, or one that has been much painted over, it may be necessary to smash it. Undo the flex connections and remove the pendant flex to the lamp. Loosen the terminals that grip the cable ends. Fit on the new base and tighten the terminal screws.

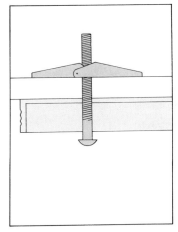

Insert the fixing screws and refix the base to the ceiling. If the ceiling is weak or damaged or the screws won't hold (important if the light fitting is heavy) then use new screws linked to an appropriate anchor fitting. The toggle plug is one used for hollow ceilings.

Renew the pendant flex and, if necessary, the lamp holder. Use heat resistant (silicone) .5 sq mm flex. Insert the flex through the rose cover and fix the wires to the terminals. An earth terminal is provided in the ceiling rose to which the earth wire is connected, if there is one. Older lighting circuits do not have them, but they are now installed for metal encased light fittings that have a three-core flex. Screw on the rose cover.

Switches

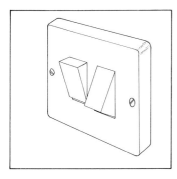

Rocker switches that you just push top or bottom to turn the light on or off are easy for the young, elderly or infirm to operate. Several switches can fit on one plate. Replacement of a switch is quite simple. Switch off at the mains first.

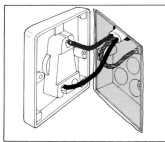

Remove the screws in the switch cover and lift the switch from the wall box. Unscrew the wire from the terminals, using a screwdriver with a fine blade. Fit the new switch by placing the wires to the correct terminals and screwing them firmly into place. The terminals are usually marked with 'COM' (common) and '1' and '2', so identification should be simple. The red conductor goes to 'COM' and the other wire to '1'. The earth wire is normally connected to the terminal in the mounting box. Screw the new switch on to the box, switch on the mains and test.

Two-way switches allow the light to be turned on or off at two points. This is useful for stairs, and also in bedrooms for the bedside and by the door. Many switches can be converted to operate 'two-way', but it needs an extra wire between the two switches—three in all.

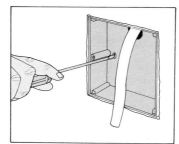

Begin by fitting a mounting box where the new switch is to go. You can use either a plastic box screwed to the surface of the wall, or a metal plaster depth box screwed into a hole cut out of the plaster.

A length of 1 sq mm three-core and earth, PVC sheathed and insulated cable is needed. The insulated wires are coloured red, blue and yellow. Feed this cable into the box through one of the knockout holes (fitted with a grommet if the box is metal).

Trim the cable, leaving enough to connect to the terminals (don't connect it yet).

Strip off about 50 mm of the outer sheathing, and 10 mm of the insulation from the ends of the three wires. Place PVC sleeving on the bare earth wire.

Next, run the cable up the wall in a channel cut into the plaster and then above the ceiling, and back down the wall to the original switch. Above the ceiling, if the cable runs in the same direction as the joists, rest it on the ceiling. If there is a chance of it being disturbed (for example in an unboarded loft) fit it to the sides of the joists with cable clips. Where the cable runs at right angles to the joists, pass it through holes drilled at least 50 mm below the top edges of the joists.

Turn off the power at the mains and remove the old switch. Feed the cable into the mounting box and prepare the cable as with the other switch.

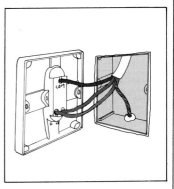

Each of the two-way switches has three terminals, normally marked COM, L1 and L2. On the new switch, connect the red wire to the COM (common) terminal, the yellow wire to terminal L1 and the blue wire to terminal L2. Place the wires in the box and screw the switch to the box.

Extending a lighting circuit

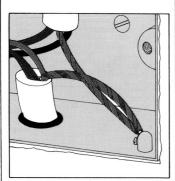

At the original switch, fit a PVC sleeve to the earth wire on the new cable and connect it to the earth point on the box, joining it with the earth from the mains cable.

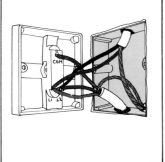

Connect the red to the COM terminal of the switch. Connect the yellow wire, together with one of the existing insulated wires, to terminal L1. Connect the blue wire, together with the remaining existing insulated wire, to terminal L2 of the switch. (The two existing wires will be red and black respectively, but either may be connected to the L1 or L2 terminals.)

Screw the switch into the mounting box, restore the power, and check everything works.

Lighting has a different kind of circuit from a ring circuit. The cable connects the lighting points and there is a branch cable from each to connect with the switch. Most homes have two lighting circuits, each protected by a 5 amp fuse. Each circuit can supply a maximum of twelve 100 watt lamps (240 volts × 5 amps = 1200 watts).

The cable is 1 sq mm two-core and earth, flat, PVC sheathed cable. In older houses the lighting circuit is often single core PVC or rubber insulated cable inside metal pipes (conduit).

The lighting circuit is wired either on the **loop-in system** or the older **junction box system,** or a mixture of both. *[Normally, extra light fittings should not be added to the older junction box lighting circuit unless expert advice has been sought.]* Study the diagrams, noting carefully how the switch forms a branch line on the live conductor.

Plan the position of the lights and the most convenient switching points. More than one switch can be operated from the same position, and a single switch can be replaced by a double switch in the same box.

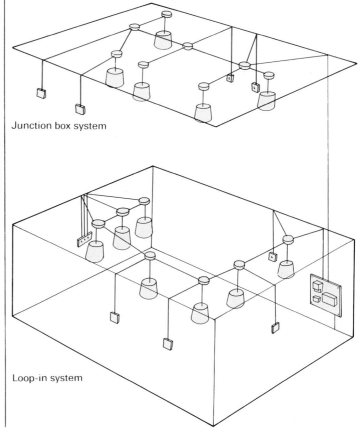

Junction box system

Loop-in system

Loop-in Ceiling Rose

A loop-in ceiling rose has live conductors connected to the centre bank of terminals. Compare this with the junction box method and make sure you understand the difference.

In a loop-in rose, the cable runs directly from the main consumer unit to the rose, which has four main terminals, each with a number of screws. Three of these terminals are linked to take the live, neutral and earth wires respectively. A wire is taken from the live terminal on the light switch and a wire from the other switch terminal returns to the four terminals on the ceiling rose. A wire is also taken from the earth terminal on the rose to the earth terminal on the switch plate or mounting box.

The *two leads* for the *lampholder* are connected to the *fourth terminal* and the *neutral terminal,* so the switch operates on the live side of the circuit. If a *metal lampholder* is connected, an earth wire must be connected to the earth terminal on the lampholder casing.

The wires to the next light fitting and switch in the circuit are taken from the *live, neutral* and *earth* terminals in the ceiling rose, continuing around the circuit until the final ceiling rose position is reached.

Mount the new switch and the cable to the switch following the same procedure as for mounting a new socket outlet (see p10), except that the mounting boxes will be only plaster depth. Take the cable from the switch to the new ceiling rose. Remove the rose carefully, marking each of the connections.

Pass a *new cable* through the hole in the ceiling alongside the

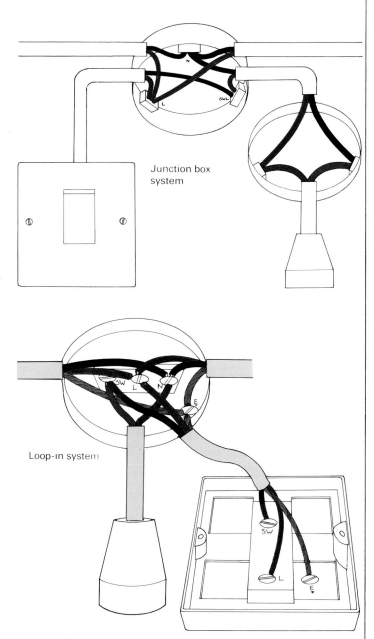

Junction box system

Loop-in system

existing cables, and take it to the position of the new rose. Prepare the ends of the cables.

At the old ceiling rose, connect the *red* wire to a *centre terminal* in the rose, the *black* to the *neutral terminal,* and the *earth* (in a green yellow sleeve) to the *earth terminal.* At the new rose, make the connections for the supply cable and switch. Refix ceiling rose and test.

BATHROOMS

No electrical heaters (or appliances such as washing machines or dryers with heating elements) are allowed in the bathroom unless they can be permanently fixed out of reach of anyone using the bath or shower. The only exception permitted in the Regs is the purpose-designed electric shower or pump. No power sockets are allowed, and no switches, except of the pull-cord type.

The only socket allowed is the special one used for shavers, and this must be the type for use in bathrooms, either on its own or with a light. These are made to British Standard (BS) 3052 and have an *isolating transformer* and earthed metal screen, so that the user is effectively isolated from the mains.

The transformer has two sets of windings so that it can be switched either to standard mains voltage or 115 volts for American or continental razors. The socket is usually made to take both the round and flat-pin plugs. These sockets are unsuitable for other appliances, and they can be wired as an extension to the bathroom light.

A room (other than a bathroom) that has a shower cubicle may have a socket, but it must be at least 2.5m from the cubicle.

A *washing machine* or *tumble dryer* cannot be sited in a bathroom, unless the room is very large and the machines can be isolated behind permanent screening (such as louvred panels) so that, effectively, they are in a room of their own.

Above *Electric fires can only be used in the bathroom if they are fixed out of reach and are of the correct design, with the obligatory pull switch. On no account be tempted to fit a power socket in the bathroom, as it is very dangerous.*

Right *This attractive and functional bathroom has a well appointed shaver socket and concealed lighting both above and below the mirror. Extra light is provided by a spotlight.*

A heated towel rail must be correctly fitted with a terminal outlet box and not an adapting socket.

Above *This well lagged tank is correctly fitted with a 20 amp double-pole switch with neon indicator.*

Left *This creative use of industrial fittings is both novel and safe. Note the water heater is also connected via conduit.*

Connecting a Bathroom Towel Rail

The simplest way to fit an electric towel rail in the bathroom is by a *spur* taken from the ring circuit to a fused, switched *connector*. Choose a connector with a red pilot light to show when it's in use and site it outside the bathroom.

From the live and neutral terminals on the connector (usually marked 'out'), take a cable through the wall to a second mounting box placed next to the heater. The heater must be mounted on the wall or floor, out of reach of anyone using the bath or shower.

Connect the flex from the heater to a *terminal block* in the box, through a surface plate with a flex entry hole. Join the flex and cable to the terminal block (including the earth wires) and then secure the block to the mounting box and screw the cover plate on.

Wall Heaters

The same method can be used for heaters mounted high on the bathroom wall and operated by a pull-cord switch. The flex is connected to a fused connector unit mounted alongside.

Wiring an Electric Shower

An electric shower is basically a water heater that heats the water rapidly as it passes over the elements. When the cold water inlet valve is opened, the pressure of the water closes an electric switch that operates the heating element. The water passes over the element and is heated—the temperature varying with the rate of flow; the slower the flow, the higher the temperature, and vice versa.

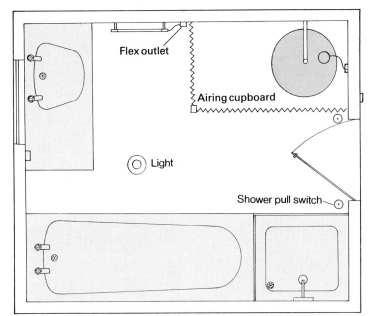

Flex outlet

Airing cupboard

Light

Shower pull switch

As the water is only briefly in contact with the element, the loading (electric power) must be high—6 kw or even higher. This means that the circuit supplying it must be 30 amps capacity and the fuse the same.

The shower unit must have its own circuit, just like the electric cooker. If there is no suitable way (spare fused circuit) on the consumer unit, it must have its own mains switch and 30 amp fuse as with cookers. (See p 13 for instructions on fitting.). This must be connected to the meter by the electricity board. A 6 sq mm cable is normally used although, if a cartridge fuse or miniature circuit breaker is used, the cable can be reduced to 4 sq mm

The *ceiling switch* must be a pull-cord, double pole (one that switches off the live and neutral connectors). It must have a 30 amp rating and a red pilot light to show when it's on. The connection from the shower unit can sometimes go direct to the switch via concealed cable, but some have a flex that needs to be connected inside a *terminal box* similar to that used for the towel rail. The terminal block must have a 30 amp rating. The pull-cord switch must be sited so it is in easy reach of those using the shower.

The route for the cable will be up the wall above the consumer unit, and through the ceiling to a point above the bathroom or shower cubicle. If the bathroom is tiled, or you don't want to redecorate, the cable from the switch to the shower unit can be enclosed in plastic trunking. Choose the shortest route, but don't sacrifice convenience. Make sure the cable is secured and follow the advice about cutting holes through joists (see p 11).

Immersion Heaters

An electric immersion heater can provide hot water throughout the year, or supplement a gas, oil, or solid fuel boiler system during the summer months, when a central heating boiler is switched off. To conserve energy (and save money) the hot cylinder must be well lagged (insulated) with a purpose-made lagging jacket at least 75 mm thick.

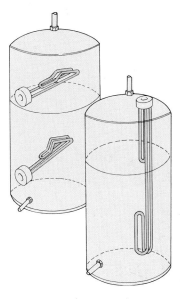

Standard immersion heaters are made in different lengths to suit different size tanks. They can be mounted vertically or horizontally. Two heaters, one near the base of the tank and one near the top, can save money. The *upper heater* is left switched on all the time to ensure a continuous (but small) supply of hot water. It just heats up the water at the top of the tank (which is drained off first). The *lower heater* can be switched on an hour or so before larger quantities are needed, eg for baths or laundry. This heater can also be controlled by a time switch.

The electricity board will connect this lower heater to their Economy Seven tariff meter, so that it uses electricity at less than half the normal price to heat the entire tank at night. This can be a real money saver for a family, and free information about how it works can be obtained from any electricity board.

The dual element immersion heater is fitted into the top of the tank, and combines the features of two heaters in one. The long element heats up the whole tank; the short element just the top part.

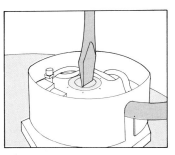

The thermostatic control on immersion heaters can be found under the metal cover on the outside casing, but switch off the heater before undoing the screw holding the cover. The thermostat can be adjusted using a small screwdriver to turn the pointer to the required temperature setting.

In soft water areas, the recommended setting is 70 degrees C, but for hard water areas the recommended setting is lower—60 degrees C. The lower setting prevents the build up of scale in the tank. Most homes find this setting sufficient for their needs, and there are savings in running costs.

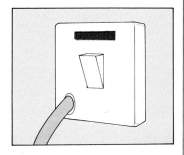

A 20 amp, double pole switch with a neon indicator is needed for immersion heaters. There are dual switches for the dual element heaters.

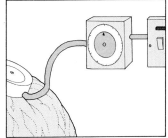

A time switch can also be fitted into the circuit, so that you can arrange for the heater to switch on and give you hot water when you arrive home. Used with a dual heater (of either kind), a time switch (without the Economy Seven tariff) may not offer big savings but it does give added convenience. Follow the maker's instructions for wiring, and switch off at the mains before installing the timer. It should be sited well clear of the hot water tank so that it won't be affected by excess heat. The flex from the switch or the clock to the immersion heater must be of the heat resistant type and 20 amps rating.

OUTDOOR WIRING

A portable cable reel connected to an indoor socket may be easy for an electric hedge trimmer but, if you want garden lights, a fountain in an ornamental pool, or you plan to make profitable use of a greenhouse, then a permanent outdoor supply is needed. The garden circuit should have its own main switch and 20 amp fuse close to the meter. An RCCB (residual current circuit breaker) should be included in the circuit.

Above left *A metal clad, switched socket outlet has been used in this workshop, complete with splash-proof switch.*

Above *A specially designed outdoor light correctly fitted on a garden wall. Remember it is unsafe to fit ordinary lights outside.*

Left *Another use for electricity in the garden is to operate pumps for ponds and pools. Here, the supply has been run from the workshop at the bottom of the garden, and the pump is concealed by greenery.*

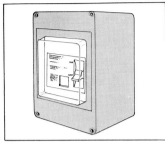

At the meter, the main switch unit with fuse and RCCB will be connected to the meter by the electricity board, once the circuit has been installed.

Underground Cables

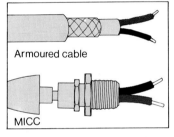

Armoured cable

MICC

Standard 2.5 sq mm twin and earthed PVC *sheathed cable*, as used for a ring circuit, can sometimes be used outdoors, provided it's carefully protected. However, it is more usual to use a special *weatherproof* one, such as 1.5 sq mm copper sheathed twin-core cable (MICC). A third type, especially suitable under a patio or for a supply to a pond, is *armoured PVC sheathed cable*. This two-core insulated cable has ribbons of galvanised steel (acting as the earth conductor) wound around the two cores and encased in an outer sheath of PVC. For both types of outdoor cable, special cable clamping glands are needed at each end to grip the metal sheathing and form the earth conection.

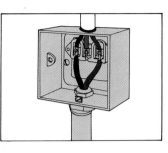

Earth continuity is essential. The outer sheath must be clamped into the terminal or mounting box. Then special clamps, that are made in sections, screw on to the cable and, after it has been inserted into the metal box, the clamp is locked into the hole by a washer and a second clamp passes over the cable and screws down on the first. This not only forms a water-tight seal, but also completes the earth link to the box. The box must then be linked to the earth conductor in the wiring circuit in the house, using any convenient terminal. The cable is buried about 500 mm below the surface. Standard 2.5 sq mm must be protected by galvanised steel conduit.

Cable Wells

If the house has solid floors, or access is difficult and conduit is being used, a 500 mm deep cable well must be dug against the outside wall of the house and the cable run into it. There should also be a cable well at the greenhouse or garage end, and both must be covered to prevent them collecting water. Different methods can be used, including filling with earth or gravel, since PVC cable is unaffected by any chemicals likely to be present in the soil.

With MICC or armoured PVC cable, the task is simpler. The only protection needed is where the cable is exposed. The cable is clipped to the wall using clips designed for the purpose. Additional protection can be given by galvanised covers screwed to the brickwork or greenhouse. Both clips and covers are sold by electrical contractors. Any covering fixed vertically must be covered to prevent water collecting, and there are metal covers and waterproof compounds made for the purpose.

In a *greenhouse* or *garden shed* the cable should terminate in a *control panel*. This must be designed for the purpose or made by an electrician, using standard electrical accessories. The component parts are a mains switch (usually with a

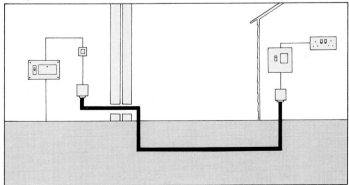

Wiring a door bell

fuse), and from this switched points and socket outlets are wired separately.

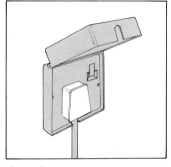

Special sockets are made for outdoor use and in protected places, such as covered car ports and patios, there are *weatherproof plastic covers*. For lighting, there are weatherproof switches.

Connector units, switches and sockets used in greenhouses and outbuildings should be installed 1200 mm from the floor. Ideally, they should be wired with MICC cable, but you can use 2.5 sq mm PVC twin and earth. The cable is taken from the outlet to each point in turn, so they are like links in a chain.

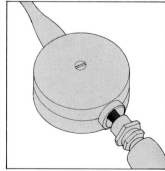

A garden pond with fountain and lights can have direct connections to a waterproof junction box concealed under a poolside stone or a covered cavity in the rockery.

A door bell or chime works on a very low voltage. This may be supplied by dry batteries, or by a small transformer that is connected to the mains and reduces 240 volts to 8, 12 or 24 volts.

A trembler bell is traditional and very reliable. A tiny amount of current causes a hammer to vibrate rapidly against a metal dome, making the familiar clear ringing tone.

A buzzer has a magnet that is vibrated by a metal rod. The pitch and volume of the buzz can be varied on most models.

A door chime uses a double-ended plunger that is mounted on a spring. This is drawn through a magnetic coil and strikes the metal chime bars to give a double note. Some models can be wired to a second push button that produces only a single note, so you can tell which door to answer.

A sonic musical door signal can be set to produce a variety of tunes, pre-programmed into the unit.

Bells and buzzers work on 3 or 4.5 volt long-life batteries. Chimes and sonics usually need 8 or 12 volt batteries. All except the sonics can be wired to a

transformer as an alternative.

A *transformer system* is needed if you want a bell push with a light, otherwise you will be constantly replacing the batteries that should normally last 3 or 4 years.

The *bell push* is simply a spring-loaded switch. When you press the button, two contacts touch to complete the circuit. The batteries are usually fitted inside the case of the bell or chime. Connection to the push is by a thin, twin-core cable called *'bell wire'*. This is sold in white, black and other colours so it can be left exposed along the edge of skirting boards, around door frames and even along the edge of coving. Tiny 'U' shaped plastic cable clips (each with their own pin) secure the wire to the surface. Take care not to sever the wire when driving the pins. The bell wire should be secured at about 300 mm intervals, with enough clips near corners to keep it tidy.

If the distance between the bell and the push is more than 9 m, you may need to increase the voltage of the batteries or the thickness of the wire (see maker's instructions). The electrical resistance of the wire

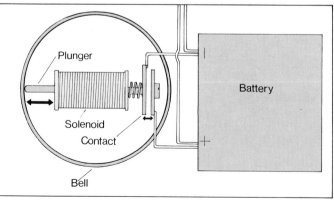

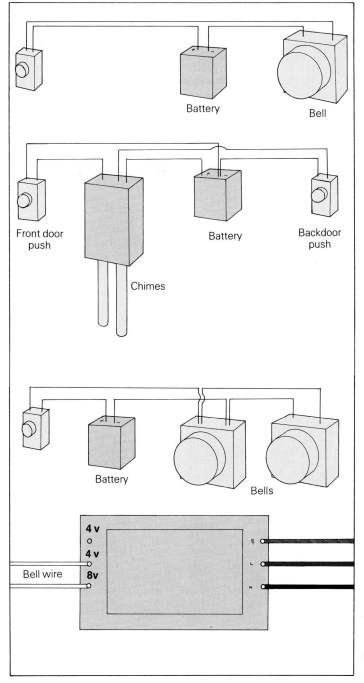

Battery

Bell

Front door
push

Battery

Backdoor
push

Chimes

Battery

Bells

4 v

4 v

Bell wire   8v

reduces the effective voltage.
There is a loss of about one-half
volt for every 9 m of wire. This
may not be noticeable when the
batteries are new, but after a
while the bell or chime will
sound weak and then stop.

It is a simple matter to have a
bell push at the back and front
doors, and some of the chimes
give a single or double note to
distinguish which push is being
pressed.

Wiring a second bell
You can also fit a second bell or
chime to work from the same
push. This is useful upstairs in
large houses, or for someone
with hearing difficulties.

Two methods can be used to
wire a second bell or chime. The
parallel method is for a pair of
matching bells or matching
chimes. The series system uses
twice the voltage, but it can't be
used for trembler bells. All the
manufacturers supply wiring
diagrams, but the basic principle
is a simple circuit that is
completed when a bell push is
pressed.

Where a *transformer* is fitted,
it must be connected to the
mains using 1 sq mm twin core
and earth cable. It can be
connected as a spur from a ring
circuit, but most consumer units
have a 5 amp fuse way designed
for this connection. If you take a
spur from the ring, use a fused
connector unit with a 3 amp fuse.

Most transformers have three
connections on the output side.
If the bell wire is taken to the
outer terminals, this gives a 12
volt supply. The middle terminal
and one outer gives 8 volts, and
the middle and other outer gives
4 volts. These combinations are
clearly marked on the case.

Top Ten Tips

1. Know where your fuse box is, and keep fuse wire, spare fuses and a torch nearby.

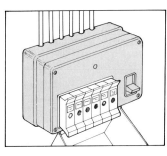

2. Label all the fuses in your fuse box, eg upstairs lights, downstairs power etc. This will save you looking at each fuse for the one that has blown.

3. If a fuse keeps blowing, there is a serious fault in the circuit or in an appliance. Disconnect likely appliances and try again. If that doesn't work, disconnect the circuit by removing the fuse. Then switch on at the mains switch to activate the other circuits. This allows you to have light for replacing the fuse or checking for the fault. Never remove or replace a fuse without first turning off the mains switch.

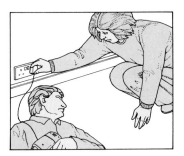

4. There is always the risk of electric shock so be sure to switch off at the mains before attempting any rewiring or electrical repairs. If you encounter any one receiving a shock the first objective must be to break the contact, switch off, remove the plug, or wrench cable free. Do not touch the person while he or she is still in contact with the electrical appliance or cable.

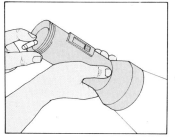

5. A cartridge fuse can be tested using a metal cased torch. Switch on the torch and unscrew the

Hall landing lights should be controlled by two-way switches at the top and bottom of the stairs.

Left *Pyjama cords tied together make an attractive pull for this bathroom light switch, and satisfy the regulations.*

base. Place one end of the fuse against the case, and the other touching the base of the battery. If the fuse is blown, the bulb won't light.

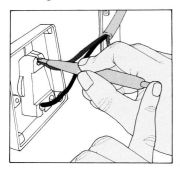

6. Some small screwdrivers have a neon tester incorporated into the handle. The neon will glow when the tip of the screwdriver is touching a live conductor and your finger is touching the metal contact on the handle. The glow is rather faint, so don't rely on it as a sure guide.

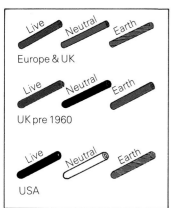

Live   Neutral   Earth
Europe & UK

Live   Neutral   Earth
UK pre 1960

Live   Neutral   Earth
USA

7. In Europe (including the UK) and many other countries, the standard colours for 3-wire flexible cords on appliances are *brown for the live wire, light blue for neutral and green/yellow for earth*. Other coloured wires may

be encountered on appliances brought into the UK, particularly those of American origin where a common arrangement is black for live, white or natural grey for neutral and green/yellow for earth. Older (pre 1960) appliances and lamps may have flexes with red for live, black for neutral and green for earth—the same colours as used for the cable wiring in the home. '

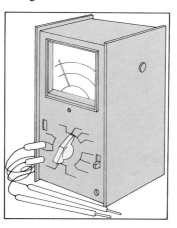

8. A multimeter is a professional, but inexpensive, instrument that is worth buying if you plan to do a lot of electrical repairs. It will measure voltage, current (amps), and resistance. Also it will check batteries and circuits.

9. It's better to add extra sockets than to use adapters for several appliances.

10. When having electrical work done professionally, always use a member of ECA (the Electrical Contractors' Association) or NICEIC (National Inspection Council for Electrical Installation Contracting).

Author
Robert Henley
Series Consultant Editor
Bob Tattersall
Editors
Dek Messecar and Alexa Stace
Design
Mike Rose and Bob Lamb
Picture Research
Liz Whiting
Illustrations
Rob Shone

Robert Henley has been writing popular articles on electricity for
twenty years. An amateur himself, who had never used a pair of
pliers or a screwdriver until he married, he has since rewired two homes.

Bob Tattersall has been a DIY journalist for over 25 years and was
editor of *Homemaker* for 16 years. He now works as a freelance
journalist and broadcaster. Regular contact with the main DIY
manufacturers keeps him up-to-date on all new products and
developments. He has written many books on various aspects of
DIY and, while he is considered 'an expert', he prefers to think of
himself as a do-it-yourselfer who happens to be a journalist.

Photographs from Elizabeth Whiting Photo Library, except for page
23 right and page 26 top left, courtesy of Tony Byers

Cover photography by Carl Warner

The *Do It! Series* was conceived, edited and designed by Elizabeth
Whiting & Associates and Robert Lamb & Company for William
Collins Sons and Co Ltd

First published 1984
Reprinted 1985 (twice), 1986, 1987 (twice), 1988

Revised edition first published 1989
Reprinted 1991
9 8 7 6 5 4 3 2 1

Published by HarperCollins Publishers Ltd

ISBN 0 00 411917 7

Printed in Hong Kong